The Explosion

"I don't really care anymore," she says in a soft but firm voice. "I don't think he loves me. He doesn't say it but he acts it. Two months ago I gave up." The wife is talking to her best friend about a relationship which is sinking fast, one which most people think is as stable as anyone would want. The husband is a success at his business and the children are doing well in school and are popular with their peers. The family has a nice home and two cars. There seems to be nothing wrong on the outside.

"What in the world are you saying?" the friend responds.

"He doesn't notice what I do or say. Hasn't for a long time." The wife is speaking barely above a whisper.

"He's much more interested in his work than in our marriage.

"He never asks me what I want to do—we only go to where he wants to go.

"What I say is not important to him.

"My marriage has become one big bore. I've gradually lost interest. I don't really care anymore." She is more specific about her feelings than she has ever been before in her life. It is as if she has been hanging onto a ledge, hoping that someone would reach down and save her, but now she has just let go. As far as she is concerned, her marriage is finished.

"I can't believe my ears," the friend says, not really wanting to believe what she is hearing.

"Is there anything I can do?"

"No, nothing."

"Have you talked to a counselor?"

"No, it's too late."

"Would you be willing to try to save your marriage?"

"I think it's hopeless. I've tried, but he just doesn't notice me enough to even realize anything is wrong. It would take a miracle. He would have to turn his life completely around. I don't think he's smart enough to know something's wrong, much less where to begin. He's so wrapped up in everything else—everything else but me. He's just not much of a husband."

"Have you told him how you feel?"

"I've tried, but he's so busy doing his things he doesn't hear me. I've been trying to tell him by the way I act toward him. But he doesn't even pick up on what I'm doing."

"Have you told him you've given up? I mean verbally told him, not leaving everything to doubt, spelling it out to him?"

"No."

"Tell him. Tell him tonight. Please. It's not fair to him to not tell him how you feel." The friend pleads with the wife—the wife of a husband who thinks he is a winner, thinks he has it all together, but doesn't realize his whole world is about to fall apart.

"I don't believe you," the husband interrupts his wife as she tries to tell him she has finally given up.

"I've been a great husband to you. You're the one who doesn't appreciate all I've done for you. Look at this house! Look at the nice clothes you're wearing! How many pairs of shoes do you have in your closet? Look at your car. I provide for you better than 99 out of 100 husbands. Don't try to blame it on me. You just aren't appreciative. I work day and night for us to be able to have what we have, to go the places we go. You just don't know when you're well off. You're spoiled and I've spoiled you. You're right. It *is* my fault. I have spoiled you."

There is absolutely zero response from the wife. There is no comeback, no arguing, no pointing out the fact that she wants *him*, not the things he has given her.

When a wife gives up, really gives up, the husband's view of the storm remains unchallenged.

After a few moments of silence that spells defeat, she finally interrupts the thick quietness.

"I am very, very tired. I'm going to bed."

She quietly, almost meekly leaves the room with her husband standing and staring at her as she walks past.

The husband sits down in disbelief. "She means it," he says to himself. "She's given up. There is no enthusiasm at all in her voice. Maybe I don't notice her enough. Maybe I'm more interested in my work than I am in her. Maybe I don't pay attention to what she says or wants to do.

"I'm in trouble," he continues out loud to himself. "Like the Bible says, 'What if I gain the whole world but lose my soul' or lose my wife? I can't believe it, but I've been blind to what I've been doing to her for years. I'm not going to take this lying down. I'm going to do something about it. If she'll believe in me and give me some time, I'll find a solution. . .somehow. I am not going to lose my wife because she thinks I'm such a poor husband."

The husband goes to bed after an hour of thinking, planning what he must do. He has run into problem situations in his business before, and he knows he has a big problem right now. He knows it is something to begin working on immediately. He will not give his wife a bunch of promises or tell her he is determined to make their marriage what it ought to be. His reputation in the business world has not been earned by what he is *going to do* but what he has *done*. He won't tell her what he is going to do. He will let her see for herself what he is doing. His plan of action will begin *now*. He plans his next day's course as he lies in bed.

He begins to picture in his mind the type of marriage he wants, and becomes determined to do all he can to develop his marriage into that picture. He decides to look for a husband who is actively and purposely doing everything he possibly can to be the best husband he can be. He will begin to look for such a husband to learn from.

The Full-Potential Husband

OVER the next few weeks he talks to all kinds of husbands: businessmen, lawyers, churchmen, athletes, schoolteachers . . . but is not satisfied with what he finds. He sees one domineering husband who "wears the pants" of the family, seemingly enjoying his role while at the same time his wife hates him. Some of this husband's friends think he is the model husband, the one to mold their lives after. But many see a flaw in the way he dictates to his wife, even though she does all he commands. He talks with this husband and asks, "What kind of husband do you think you are?"

"I'm a firm husband who runs the family the way it ought to be," he is informed. "An authoritative husband. I take charge." He keeps hearing one word in the conversation: *pride*.

The man continues to talk with husbands who appear to get along with their wives very well. There is a lot of smiling, opening doors, sending flowers, and buying new cars. Whatever the wife wants, the husband obliges.

Some of these husbands think they are good husbands. Some even boast that they send their wives flowers every week. He hears them build themselves up as they talk about how good they are to their wives. They like to tell other people about all they do for their wives. But are they really acting out of deep love, or are these simply superficial attempts to please their wives? Are these husbands really only fooling themselves?

The more husbands he interviews, the more he realizes that most are basically self-centered. The authoritative husbands are thinking directly of self in the husband-wife relationship. The "ideal" husbands are really thinking indirectly of self most of the time because their actions are done more out of pride than out of love. They enjoy telling other husbands how "ideal" they are; seldom do they do anything secretly or quietly for their wives, without any fanfare attached.

With neither type of husband does he find a man reaching his full potential as a husband. There is too much missing of what he is looking for. He knows exactly what he wants to find, and commits himself to keep looking until he finds the full-potential husband, one who is working toward being the best husband he can be. Then he will begin to learn from such a husband and start to mold his relationship with his wife that will lead to the marriage he has pictured in his mind—a happy, loving, fun marriage, one in which he will look forward to going home rather than working that extra hour, making one more halfhearted call on a client, or watching television rather than talking with his mate. He visualizes a warm relationship built on love for each other. He knows that if he can develop into the right husband, his wife will respond to him warmly.

She has given up on him, but he has one more chance to pull it together again. If he fails this time, it will be all over. He isn't going to take advice from any superficial, shallow-based husband. He is going to look until he finds a husband who means business, who knows what the bottom line is all about.

He continues to look, and finds a few husbands who almost have what he is looking for, but there always seems to be something missing that keeps their marriage from working at full potential.

One day a friend tells him about a businessman who sounds a lot like what he is looking for. His wife respects him, loves him, and supports him, and he treats her like a queen. He takes her out on dates, but when he does this or something else special for her, he doesn't make a big deal about it to anyone. He even does some things for his wife that she doesn't know about herself, and some things she will never know about. These he does just to reinforce to himself the fact that he loves her.

"He sounds like the husband I've been wanting to interview. How do I get in touch with him?" the man asks his friend.

He wonders if it is really true, and whether this man would be willing to share his secrets of being a successful husband.

It turns out that this special husband lives only an hour's flight from the man. He decides to call him and see if he will give him an interview.

"May I tell him who is calling?" the secretary asks, and then puts the call through.

"I'll be happy to spend some time with you," the businessman responds, "but only on one condition: You must be sold out on the idea of becoming the best husband you possibly can. If you will give me that commitment, I will arrange to meet you at my office on Friday."

The man begins thinking after hanging up the phone, "I am determined to see this through. The answers to my curiosity about what he has to say and what secrets he knows that I don't know—if he is really genuine—will be worth the price of the plane ticket."

The husband tells his wife he has a very important call he needs to make personally in a nearby city. So far nothing has changed at home except that he is treating her better than before and showing her more attention. But she sees through all that. She knows his actions have changed a little but his heart has not made a basic change. It will only be a matter of time until he is back in his busy rat race again. She has seen it happen so many times before.

"That's fine—I don't mind," the wife responds to the husband. She doesn't mind if he goes out of town. She doesn't mind if he *stays* out of town. She feels so badly about her marriage and her whole life that she can't analyze how she feels about him without having it affect her whole life.

The husband begins to sense what she is going through and becomes more determined to meet this biggest challenge of his life head-on.

After the short flight he arrives at the office and is met by a gentleman dressed in a business suit, neat-looking and with the kindest, most-relaxed face he can ever remember seeing.

"Good morning, I'm Ron Farris." The gentleman quickly walks from behind his mahogany desk and reaches out his hand. "Sit down and tell me how I can help you."

"I understand you're a perfect husband," the man responds. "I want to improve my marriage, and I thought you might share some of your secrets for being a better husband."

"I'm not a perfect husband by any stretch of the imagination," laughs the gentleman. "But I'm one of the few men I know who works at becoming a good husband. You see, the secret of a good marriage rests with the husband. If you reach your full potential as husband, you'll build such a respect from your wife that she will begin to build the type of relationship you're looking for."

"That sounds great," responds the man. "You say you're one of the few who work at this plan for husbands. You mean there are more?"

"There are five of us in our group," the businessman responds. "Of course, there may be many such groups forming by now. We have told several interested men like yourself about what to do. Whether they follow through with anything is completely up to them. It takes different types of commitment from each individual husband, and it's something you have to commit to and do in your own special way."

"This is interesting," responds the man. "Do you five have an organization or a name?"

"Not really," he answers. "We each call ourselves 'the full-potential husband' because that is our goal and we work at it daily. Each of us has read a book entitled *Reaching Your Full Potential*. It's a book about setting goals to reach your full potential in different aspects of your life. It so happens that the five of us reached the same conclusion about our lives. None of us had ever set a goal to be the best possible husband we could be. We had set goals in business but never in our spiritual lives or our families' lives. We began to meet on a regular basis and discuss what kind of goals we should set to develop into the type of husband we should be; we began to call ourselves 'the full-potential husband' because we are striving for true intimacy in our marriages."

"The full-potential husband," the man responds. "That says it all. That's what I've been trying to become. I want to reach my full potential as a husband. How do I do it?"

"It will cost a lot," Mr. Farris says.

"How much?" the man asks.

"A lot."

"Do you think I can afford it?"

"Yes, any husband can afford it."

"But didn't you say it would cost a lot?"

"Yes. More than most husbands want to commit to."

The man thinks a minute. "Okay. I'll go with it."

Mr. Farris takes out a pad of paper.

*

It's Never
Too Late
To Begin
A
Great Marriage

*

"I'll make a list for you. What's your favorite hobby?"

"Golf."

"Good." He writes that down and asks, "How do you spend your time in the evenings?"

"I read. Watch some TV."

"What material possession do you want most right now?"

"I'm an avid jogger. I'm planning to buy a new pair of jogging shoes—the expensive kind."

"What do you dislike about your wife the most? How does she really upset you?"

"She's late. Always late. We always end up arguing by the time we get anywhere because we're late and it's her fault."

"If you had your choice, where would you take your next vacation?"

"Mid Pines Golf Hotel. There's a lot for the whole family to do there."

"If your wife had her choice, where would she take the next vacation?" He continues to make notes on his pad.

"Oh, the beach. But that's such a drag."

"Do you ever give your wife a gift other than at Christmas?"

"Well, I sent her flowers on Valentine's Day."

"Anything bigger than flowers?"

"Not really. I provide money for clothes. She buys a lot of clothes. Lots of shoes, too."

"Now look at me and listen," Mr. Farris exclaims as he glances at the list he has just made. "Your golf, your reading, TV time, your jogging shoes, your dislike of your wife's tardiness, the Golf Hotel, an expensive gift—all this and much more may be the price of your becoming a full-potential husband. If you're willing to buy the package, then I will talk seriously about helping you get there."

"But that leaves me with nothing I can consider mine."

"Exactly. Do you still want to join the ranks of a full-potential husband?"

There is a slight pause and silence. Then the husband responds, "Okay, I'm willing to pay the price if you're sure it will better my marriage and make a fuller, richer life for me and my wife. You see, she's at the point of giving up on me, and I've got to do something that's guaranteed to work. I love my wife and I don't want to lose her."

"The basic difference between what you *have been doing* and what you *need to do* is the difference between 'I love you' and 'I love you very much.' It comes down to caring about how you feel about your wife. A wife can tell if you really care about what she thinks, what she likes, about *her* side of life. She is either satisfied with you as a husband or she isn't."

"Well, I can tell you she isn't," responds the man.

"Is your concern more for yourself, or are you really trying to make your wife's life more pleasant?" Mr. Farris continues. "It's not something you just try to do one day; it's a philosophy you have to develop each day."

"You mean if I start being more concerned about my wife's feelings, then she will respond to that feeling in a positive way?"

"Exactly. Sooner or later she will respond if your love is real," Mr. Farris states with certainty.

The man begins to remember a part of himself he had long forgotten. He begins to think of his wife more like when they first dated—when the little things didn't bother him.

Mr. Farris interrupts the man's thoughts. "Try walking in her shoes. Try seeing it from her vantage point. Help her to reach her needs. Take some interest in what she is interested in. Begin to see this working into an exceptionally exciting and enjoyable relationship. Women just love being the wives of full-potential husbands. Think about that five-star relationship you want. See it in your mind. Set it as your major goal. Then a lot of the rough edges will be smoothed out. If you see that relationship long enough, a strange thing will happen: The goal picture will become real."

"This all sounds good on paper, but does it really work?"

"It works. I wouldn't be spending this time with you if it didn't. Anytime you are reaching your full potential, your life takes on new meaning," Mr. Farris continues. "It becomes exciting just going through a normal day. You won't be sorry you made this decision. That promise I can make unequivocally." He picks up his phone and makes a call. He writes down some dates as he talks and hangs up.

"Here are four dates for you to choose from. Dr. John Parrish will meet with you on any one of these days. His phone number is written with the dates. Give him a call when you decide."

"Why are you going to all this trouble for me?" The man takes the paper with the dates and puts it in his pocket.

"Because I know a secret that works and I want other people to know it too. I once was in the same shoes you're in. I wanted a better life, and I knew it had to start with me and my relationship with my wife. And, like you, I realized that one of the bases for reaching the full potential in my business was somehow related to reaching my full potential in my relationship with my wife. I admire any man who wants to reach his full potential. I'll help in any way I can."

"Then why don't you speak to groups of men and tell them about it?"

"Simple. If a man isn't interested enough to fully commit to it, to buy a few airline tickets, to take some time to investigate, to decide to go all-out to become a better husband, then it would do no good to tell him how to become a full-potential husband. I don't have the time to waste on him. I admire you for your interest in your marriage. It's my desire that you become a full-potential husband yourself someday."

The man rises and extends his hand. "Thanks so much for your time and interest. I'll always be indebted to you."

He can hardly wait to get back to his office and check his calendar to see when he can visit the next full-potential husband. Already he is feeling an indescribable satisfaction and peace building up inside. He knows that these men who call themselves full-potential husbands mean business and that he will benefit from their expertise.

As he goes from his office to his home he rehearses in his mind what he has learned from the first full-potential husband. He sums it up in one phrase: *commitment to a goal*. He has made the decision to commit to the goal of becoming a full-potential husband, to reach his full potential as a husband.

The strangest sensation overtakes him as he walks into his house that evening. He sees his wife in a completely different light for the first time. Even though he talks with her but doesn't get much response, it doesn't bother him so much because he knows something she doesn't: He knows he has committed to being a different husband than before. He knows it will only be a matter of time until she starts responding to his love. His actions around her are no longer to try to please her but to love her.

His wife knows he has been out of town all day but doesn't care enough about it to even ask how his day went. She doesn't notice any difference in his light kiss he gives her when he comes in. He has done that before, but she knows it doesn't mean a thing. She has not realized her true feelings toward him. She doesn't realize that there is a very fine line between "don't care" and "hate." She does hate him in a lot of ways, but mostly she doesn't care.

She has never considered divorce before. Her parents had taught her that divorce was not an alternative to a troubled marriage. But she is beginning to contemplate it more as time passes. It's strange to her how her resentment is building up toward her husband. First there are the thoughts—thoughts she never allowed herself to think before, thoughts of dislike in little mannerisms of her husband. She doesn't even like the way he walks. His smiles seem more like smirks than anything else. She used to like the way he came home with his top button unbuttoned and his tie slightly loose, but now it's irritating to her. Who does he think he's trying to look like? Some movie star? He thinks he's so hot with his good job and nice car and all that goes with it. Well, it's not impressing her one bit anymore.

She thinks she will spend the summer with her mother. That's acceptable. Her friends would accept that as an okay thing to do with the children. At least it will get her out of the house and away from him for awhile. She needs some time to think her life out. She is having an awful time, and it's getting easier and easier to blame him. He thinks he's never wrong, but this is one time he's going to see how wrong he's been, and it's going to be too late for him to do anything about it.

Commitment to a Goal

TWO weeks later she doesn't care when he tells her is is going out of town again.

He finds himself at a physician's office, supposedly at closing time. However, there are still two patients in the waiting room. Thirty minutes later he is ushered into the doctor's office.

"Sorry about the delay. Can I get you a cup of coffee?" Dr. Parrish looks relaxed even though he is between office hours and evening rounds at the hospital.

"No, thank you. I really appreciate your time. I know how busy you are."

"That's why our full-potential husband group gives me this particular assignment," laughs the doctor. "It's short and to the point. You'll be on your way back home in 45 minutes, and I use this time as an excellent break in my routine."

"If it's okay with you, we'll get right down to business."

"Fine with me," responds the man. "I have my pencil sharp and I'm ready to learn."

"Good. I've already learned about your particular situation and have gone over your resume you sent last week. I'm going to make some assumptions; if I'm wrong, tell me.

"You consider yourself a fairly hard worker; you do what is expected of you at work; you attend church fairly frequently; you say you love your wife but are about to lose your marriage. Is that a fair assessment?"

"Close enough," responds the man.

"Good," he continues. "You are seeking ways to become a better husband, to perhaps save your marriage. You don't really think you're doing all that much wrong, but neither do you think you're doing all that much right, or else your marriage would be more successful."

"That's about the way it is," the man responds again.

"A recent survey shows what characteristics most wives want in a husband." The doctor pulls out a sheet of paper from his desk drawer. "Some of the qualities you have. Many you don't."

"Thanks for being candid."

''We'll waste less time if I'm honest up-front with you. I'll tell you the facts; you take them for what they're worth.'' The doctor is pulling no punches. He's putting it on the table, take-it-or-leave-it style.

''What did the survey show?'' the husband wants to know.

''A wife wants a husband who listens, is understanding, is confident in himself, has security in his job, is dependable, is an achiever, is aggressive but with humility, and is trying to make the marriage better.''

''I see what you mean. A lot of those things I'm not, but at least I'm trying to make the marriage better.'' The husband looks up from the notes he is taking.

''Let me go over some more facts,'' the doctor continues. ''Statistics show that the national average for divorce is about one out of every two marriages. But where three things are done regularly, the divorce rate drops to one in 400.''

''What three things?'' the husband is eager to know.

''Regular Bible reading with husband and wife, regular prayer together, and regular attendance at church together.''

"That sounds like pretty good marriage insurance! But some of that I don't think will interest my wife very much."

"Are you willing yourself?" The doctor looks at him sternly. "In the world of today, I can't give much hope to a marriage without both husband and wife relying on God's Word, the Bible, to build their marriage. All other advice you may get simply doesn't hold up. The odds of one out of two just doesn't compare to one out of 400."

"I'll try anything," the man says, desperate to find some answers that really work.

"Okay. But before it can work for your marriage, you must first get your own life in order."

"Such as?" responds the husband.

"You must see yourself in a strong light, and your overall life must change. The way you are handling your work and your spiritual life must change."

"Wait a minute. You're comparing my work and my spiritual life."

"They go hand-in-hand. One intertwines with the other. You can't separate the two any more than you can separate husband and wife from marriage."

"In that case I do need to work on my whole outlook." The man begins to adjust his reasoning about his overall viewpoint of what he should be as a husband. He begins to realize that it includes how he thinks of himself and his work as well as how he treats his wife.

The man continues, "How do I go about making this improvement?"

"Do you set goals for yourself?"

"Not really. I get things done, but I don't actually sit down and write out a particular goal. If I'm productive, which I am, is it that important to set goals?" The man is going to keep the full-potential husband honest in any advice he offers. That's the way both men want to play the game.

"Look. Every man has within him certain qualities, abilities, and talents that no other man has. What guarantees that these will ever be found? What will make them evident and usable? What will pull the most out of a man and stimulate him to obtain his greatest productivity, his full potential?"

"You're saying goals are all that important?"

"They're everything!" exclaims the doctor. "They're the difference between mediocrity and full potential. One of the saddest stories ever told is 'I could have been' or 'I could have done'—the story of the man who didn't live up to his potential in life. That man reaches middle life and looks back in remorse, then continues on in his sea of mediocrity, never realizing the importance of setting goals to become the individual that God planned for him to be."

*

A Truly Great Marriage Is An Obtainable Goal

*

"You say setting goals for your work, your life, and your spiritual growth overlap?" The husband is really set back by all this. He realizes he may have been missing out on a lot in life.

"Yes and no," the full-potential husband responds. "Goals can be made for self alone, but if that is the case, the accomplishments and success which they attain are shallow, hollow, and empty. The only true success comes when your goals are God-directed, Holy Spirit-empowered. The goal of such a man has to be directed toward being God's man, a man God can use and rely on.

Man is made up of more than what the world thinks and sees as a successful man. Every man has a spiritual vacuum placed within him by God. Unless that vacuum is filled by direction of the Holy Spirit, that man will never feel success. Oh, he may obtain wealth, and look successful by most who see him, but I guarantee that if you get to know him, you will realize that he is not truly successful. He will admit to you that there is something missing. He will tell you that the new car didn't fill the expectations he had for it. He will tell you that the vacation wasn't as much fun as he thought it would be. He will tell you that there is inner peace missing. Watch him when he reaches a certain age in life. Sooner or later he will come onto something that will throw him. He will not be able to handle the situation because he has not developed a reliance on God. That inner void is too great; he relies only on self, and eventually comes tumbling down."

"I wouldn't like to end up like that," the man says.

"Look at Scripture. Who does God use? What type of man does He use?"

The man hesitates for a moment.

Dr. Parrish continues, "A man of action. Look at Moses, Paul, David, and Daniel. These were men of action. You need to set some goals and get on with it. Your whole life will change, and your outlook on yourself will change as well. Your outlook on your work will change and even your outlook on your wife will change." The doctor is so excited that he has to refrain from sounding like he's preaching with only one person in the congregation.

"Thanks. It was stimulating to hear that coming from you. You've made your point. Now, how does this relate to me and my work and my wife?" The man is now accepting the plan 100 percent.

"Your major goal in life has to be directed toward becoming God's man and being a living witness for Him. It will change your work and it will change your marriage, all for the better."

"How?"

"One of the greatest principles found in the Bible is found in Genesis. Do you recall what happened when Abraham decided it was time for his son Isaac to have a wife?"

"Not exactly. Go ahead and tell me."
about it."

"Well, Abraham sent his servant to his home country to look for a wife for Isaac. The servant prayed to God to let the woman He wanted to be Isaac's wife be the woman who did more than was expected of her. When he asked a girl named Rebekah to draw him some water, she said she would not only draw him some water but would water his camels also. She became Isaac's wife because she did more than was expected of her."

"A good principle," the man agrees.

"One of the greatest principles ever written. Can you imagine what would happen in your work if you started doing more than was expected of you? Can you imagine what would happen to anyone who consistently did more than was expected of him, who got to work 15 minutes early, stayed 15 minutes late, and did more than just his job description?"

"Sure," the man replies. "That person would get a raise and an advance in job position."

"That's true. You would advance toward the top and get raises and bonuses. But that's not the real reason this principle is so important. The real reason is found in Jesus' teaching in Matthew and Paul's teaching in Ephesians. Jesus says that if a man makes you go one mile, you are to go with him two—do more than is expected. He also says that if a man asks for your coat, give him your cloak also—do more than is expected. Paul instructs Christian slaves to be the best slaves possible, to work as if they were working for the Lord and to do more than is expected of them."

"I'm beginning to see your point."

"Jesus and Paul were giving instructions in this 'do more than is expected of you' principle for one purpose only: so that you may be a better witness to those around you."

"I see what you're saying," the man responds.

"And this includes your wife. Let her realize that you're serious about changing yourself toward a more godly man and she will better understand your attempt to change as a husband."

"And that's where the Bible reading, prayer, and church attendance with her comes into play?"

"Exactly. You can't isolate just wanting to be a better husband from wanting to be a better man and doing better on your job."

"And the spiritual emphasis in your life pulls it all together?" The man quickly develops insight into what he must do in the days ahead.

"It is the matrix, the common denominator, the mortar which allows you to build and develop into your fullest being.

"Now what are you going to do about it?" The doctor is almost through. He wants the man to make some decisions right now.

"I can begin my own spiritual growth by reading the Bible," he begins.

"Study. Don't just read. *Study* the Bible. What else?"

"I can start looking at my job as a basis for witness for the Lord!"

"Good. Don't be afraid to work. Work like you are doing it for Him. Do more than expected. It will be noticed and will become an excellent opportunity to witness to those around you who respect you."

"And all of this is going to make me a better husband?" The man hasn't lost perspective of why he came here today.

"You will be a better *you* because you will be Christ-centered, not self-centered. You will become a better employee because your work will be based on God's Word. And you will eventually become the husband you should be because you will have developed a foundation which will stand all tests." The full-potential husband shuffles a few charts on his desk, and the man realizes that the doctor is finished and ready to make rounds at the hospital.

*

I Am What
I Am
But I Can Change

*

"Thank you so much." The man reaches out his hand to the doctor. "Thank you for giving me a glimpse of what it can be like to get out of this sea of mediocrity."

"You're welcome. Now go and get started on some goals in your life!"

The man returns home and reviews his visit with the doctor. For the first time in his life he writes out some goals for himself, his job, and his relationship with his wife.

He decides to order some Christian magazines for his office waiting room. He has been told about a course put out by a group called The Navigators. They furnish an organized program for memorizing scripture. He decides to order the program and begin memorizing verses. He also decided to order some Christian books to give away to his clients. He commits to spiritual growth and becoming a witness for the Lord in whatever little way he can.

His wife notices some changes around the house but is not sure what they mean. She saw her husband reading the Bible instead of watching television last night. He is leaving for work a little earlier than usual but seems more organized with his time. He actually seems to have more time at home. He somehow seems more aggressive but at the same time appears to be developing humility.

There is one significant change she *can* see that may make a difference in the marriage. He wants to have a time of reading the Bible together at night. That in itself seems like a worthy plan for him to come up with but the most awkward situation he has put them in has been his suggestion of praying together. She just finds that very difficult to do. She has prayed before but not out loud. She has never liked the idea of praying aloud in public, much less in front of her husband. It's *his* idea; she will let *him* do the reading of the Bible and the praying.

One thing she hasn't noticed, though, is any significant change in their marriage. It isn't getting any worse, but it's not getting better either. She begins to admit to herself that living with a roommate called "husband" isn't a very pleasant way to live. She can't continue on and on like this. It is just a matter of time before she must consider a separation or even divorce. She realizes that her husband is honestly trying to be a better husband, but too much resentment has built up—too much bitterness for his changing to ever help now. Why didn't he do something when she still had some love in her heart for him? She is angry at him now because of his timing. All his Bible reading and self-improvement she sees him doing mean very little to her now. She almost feels sorry for him. . .too little, too late.

The husband does not sense that the marriage has deteriorated to this degree, and he pushes on the best he knows how. He is not centering his efforts so much on getting back with his wife as he is on getting his own life in order. In fact he centers most of his effort on setting goals. He sets no goals without first bathing them in prayer, as the doctor had instructed. He writes down his goals and commits to them wholeheartedly. He becomes determined to persist until his goals, under the direction of the Holy Spirit, are reached. He sets goals for spiritual growth, prayer time, and Bible study. He sets goals related to his job, planning how to do more than is expected of him. He sets goals for his relationship with his wife. Perhaps it is best that he does not know his wife's true feelings about him.

Overall, the man has noticed a basic change of attitude in his life—not just toward his wife, but in his work, in how he deals with other people, and in how he views himself. He is becoming a more humble man, but he feels more confidence in himself. For the first time he feels an assurance that he is going to make it. Things are going to work out with his wife.

He is ready for another learning session as he prepares to meet the next full-potential husband, Lowell Benjamin.

*

The Saddest Words
In Any
Marriage Are
"I Wish I Had..."

*

A Return on Your Investment

LOWELL Benjamin is a stockbroker and investment counselor. As the man walks into Lowell's office, he is met by a man in his mid-forties dressed in a dark-blue pin-striped suit, white shirt, silk tie, and polished shoes. Anyone can tell that Lowell is a businessman who takes his work seriously. But behind this external appearance is a warm smile and a certain gentlemanly humility and likability.

"Welcome. I've been looking forward to seeing you ever since I heard you had decided to try the plan. It's a big price, but worth it." The full-potential husband is quite sure of himself. If it didn't work, he surely wouldn't be giving up his valuable time for free.

"Thank you. I'm ready to listen and learn."

"Well, I'm an investor and I'm going to talk to you about an investment. I call it Wife-Time investment."

"Wife-Time investment. What's that?"

"It's the third of five steps to becoming a full-potential husband. If you can't do this one, there's no need to go any further. You learned the reasoning of commitment from the first full-potential husband, and the importance of setting goals from the second."

"Five steps? I didn't know there was a progression to becoming a full-potential husband," the man says.

"Yes. It's a growing situation, just like anything else in life that's worthwhile. Nothing great ever happens instantly."

*

A Marriage Needs
Attention
Like A Garden
To Keep
The Weeds
Out

*

"Yes, I know. I tried instant grits once. I thoroughly understand your point." The man laughs. He is already liking and respecting this new full-potential husband.

"Just how does Wife-Time investment work?"

"It's really very simple to understand. The difficult part is committing to do it. But you have already committed yourself to this through to the end, so it should be relatively simple for you.

"It's just a matter of investing lots of time with your wife. Do you spend lots of time with her?"

"Oh, I don't have that much time to invest 'lots' of time with her. But I make up for it with quality time spent with her."

"Quality time?"

"Yes, you know—take her to this golf hotel for a week. We eat out every evening. She doesn't have to worry about cooking or washing dishes or anything like that for a whole week."

"Does she play golf?"

"No, but she will occasionally ride with me in the golf cart just to be with me and get outside for awhile. She really enjoys the natural beauty of the course and the clear air. She likes that kind of stuff."

"When was the last time she rode the course with you?"

The man thinks a minute. "Two years ago she rode the front nine with me."

"The front nine?"

"Yes. She had to wash her hair or something like that or she would have gone the full 18 holes."

"And that is quality time with your wife?" Mr. Benjamin writes some notes on a pad.

"Go on. Tell me more quality time you spend with your wife."

"I take her to a very nice restaurant on our anniversary. Plus we go out for dinner often."

"Go on."

"Well, I took her with me on a business trip last year. One of the best times we've ever had together."

"Whose idea was that?"

"Actually, the company encouraged it and paid the difference in airfare and room. We really enjoyed that trip."

"What else?"

"I'm sure there are other times when we spend quality time together; I just can't think of them offhand."

"Okay. Look—quality time with your wife is important, but the time you are calling quality is centered around *you*, not your wife.

"I understand that you are a golfer and want to take your vacation at a golf hotel. But I also understand that your wife likes the beach. Have you considered taking your next vacation at the beach?"

"There's just not that much to do at the beach."

"There's not much to do except spend time with your wife for a week—actually sitting in the sun and talking with her, walking down the beach for an hour at sunset, reading parts of a book out loud to her, rubbing her suntan lotion on her, joking with her, taking in a movie—her choice—and soaking up some sun together."

"But I can't take the sun well. My doctor says I'm prone to skin cancer."

"That makes it a little tougher in your case. Can you think of anything you could do to get around that problem?"

The man thinks a minute.

Mr. Benjamin remains silent and waits.

The man slowly begins to nod his head and smiles. "I think I'm beginning to see a little bit of what you full-potential husbands are all about. If I want to spend quality time with my wife, all I have to do is commit to do it. I know what you would tell me to do—get a beach umbrella and sit in the shade while she sits in the sun."

"Exactly. It's going to be fun working with you. You already have good basic insight. Now, how would you handle the other 'quality times' you mentioned? How about the restaurants you take her to on those special occasions? Do you take her to *your* favorite one or hers?"

"Well, the one I usually take her to has both seafood and excellent steaks. She likes seafood and I like steak. We're both satisfied."

"Is there a seafood restaurant in town?"

"Oh, yes a very excellent one, but their steaks leave a little room for improvement."

"If you really wanted to treat her to a good quality-time night—just for her—what could you do on your next evening out?"

"You know, I really get enough good steak with business associates to last me a lifetime. I see what you're getting at: Center quality time around her completely. I'll take her to the seafood restaurant."

"Right," responds Mr. Benjamin. "I don't mean you have to give up ever eating where you want to, but if you are going to call an evening 'quality time' with your wife, then get self out of the picture and give her your individual attention that particular evening.

"Now, what about that business trip you all had such a good time on? Could you take her with you more often?"

*

Invest In Your Wife— Spend Time

*

"I'm not sure. That was the first time my company has ever sent our wives with us to a meeting. I haven't heard them mention doing it again."

"Are you saying," Mr. Benjamin leans forward in his chair and raises his eyebrows just a little, "if your company never sends your wife with you on a trip again, you will never take her with you again?"

"It costs money for her to go, I hope you realize. I don't have that kind of money," the man states authoritatively.

"Are you aware of special airfares that allow you to take your wife with you at a very minimal addition, and that most hotels charge very little or nothing extra for your wife?"

"I've heard that advertised but never checked into it. If I set my mind to it, I could probably forgo that golf jacket and new putter I was about to buy. Yes, we could cut down on several things we have planned to buy and she could go with me. I'll admit it would be an inexpensive way for us to spend some quality time together."

"Good." The full-potential husband leans back in his chair and relaxes. He is very satisfied in how the man is coming along.

"So much for quality-time investments," he continues. "Now what about *quantity*-time investments? If I'm going to advise you in investments, I want to cover it completely.

"What other time do you spend with your wife?"

"You mean like every day?"

"Like every day and night."

"Well, let's see—this is a hard one. Since I have to work so hard at my job to provide for my wife and family, we don't really spend a lot of time together except on those special occasions I mentioned earlier. Some of the time she has eaten dinner by the time I get home in the evening. Plus, I have meetings three nights one week and two nights every other week. We hardly see each other those days."

The full-potential husband continues with his notes and asks, "Do you watch television together?"

"Sometimes."

"Do you talk to each other while you watch TV?"

"Not really. You can't talk and watch too."

"How often do you watch?"

"Not much at all. Maybe a couple of shows a week—an hour-and-a-half, maybe two hours a week. Plus I watch the news when I make it home on time. Sometimes I watch the late news or the first half-hour of the Johnny Carson Show. That's the funniest part."

Mr. Benjamin makes more notes, adding up the time the man invests in watching television.

"On any investment I advise, the primary goal is return on investment. What return do you get on the time you invest watching TV?"

"Well, I . . ." the man hesitates. "I enjoy watching some things. But I admit, a lot of it's just time spent. I never thought of it as an investment I needed to make a return on. That's good—time really is an investment!"

Mr. Benjamin interrupts. "You've said you don't have much time to spend with your wife. The best investment you can make to improve your relationship with your wife is to spend time with her. Tell me, what is your wife doing at home while you watch the news, read the paper, or watch Monday Night Football or the Johnny Carson Show?"

"Oh, just the usual—cooking dinner when I first get home, or she may look at a magazine while I watch a ball game. She spends at least 30 to 40 minutes just getting ready for bed."

"Take just the facts we've talked about so far. How could you invest some quantity time with your wife? Remember, *quantity* time is more important than *quality* time in the overall wife-time investment because it's repetitious. Repetition gets the point across: You think more of being with your wife than reading the paper or watching television or doing anything else with a low return on your investment in time. How could you invest this time differently?" Mr. Benjamin folds his hands and waits in silence once again.

"I believe I can think this through, although I'm not sure I could invest time with her while she's cooking dinner." The man is trying to adapt to the idea of wife-time investment.

*

WIFE
Top Priority

*

"Can you do anything to help her?"

"Not usually. I can't even boil water right. I could possibly put ice in the glasses or pour the coffee."

"Set the table?" the full-potential husband encourages.

"Yes, I know how to set the table."

"Stir the vegetables?"

"I could do that."

"Put food in the serving dishes and take them to the table?"

"Sure, I could handle that. You've helped your wife at dinner, haven't you?"

"If you wanted to make a significant investment in time with your wife, and with it get a greater return on your investment, what else could you do?"

The man thinks back over his time at home that he could invest. "I could spend some time talking to her. Even while she's cooking, I could talk to her."

The investment counselor eggs him on: "Talk to her about her day? About the children? About anything she is interested in?"

"Yes," responds the man. "I could invest some time talking to her while she's getting ready for bed. That would be a lot of time invested—a whole lot of time."

"What else?"

"I have to drive back to the office occasionally to pick up something. I could take her with me. That's 40 to 50 minutes round trip."

"Do you ever take her for dessert or coffee on Monday night? You know—while the football game is on?"

"Never have, but I could." The man is smiling broadly now. "Okay, I've got it now. I never thought I had any time to spend with my wife, but when you look at it as an investment, I've got all kinds of time. And you say quantity time is all that important?"

"Very much so," responds the broker. "If you take the wife-time investment concept seriously, you will be amazed at the response you receive. Show her how important you think she is. Give her something money can't buy—your time. She's worth more than money, you know." Mr. Benjamin points to a small plaque under a wood-framed picture of his wife:

SHE IS WORTH
FAR MORE THAN RUBIES
Proverbs 31:10

The man leaves the office and heads home eager to begin working on his wife-time investment. As he heads for the airport, he thinks back about how simple wife-time investment is, especially in comparison to the returns he will receive from the amount of time invested.

He begins to get excited as he heads home to apply some of what he has learned. He realizes that before he can take her somewhere for some quality time he has to spend some basic *quantity* time with her. He will start immediately. Normally he would go by the office and work for an hour or two, but today he makes it home by five o'clock.

In one short week, for the first time in ages the wife notices something different going on. Her husband has made it home at a reasonable time for over a week now. He has missed two of his night meetings, stating that they weren't covering anything very important and that he didn't need to go. She also notices that he actually asks her what she has done that day and what she is planning for tomorrow.

He has begun to put ice in the glasses and pour the coffee for dinner. He asks her to ride back to the office with him one evening and even asks her out to lunch one day. But she declines; she doesn't really feel like riding to his office or having lunch with him. However, she does admit to herself that she senses her husband is paying more attention to her. It may be her imagination, but he seems more genuinely interested in her and there doesn't seem to be quite the tension between them as before. For the first time in a long time she actually enjoys telling him something about the children.

It is a relaxed evening. They are both sitting in the den glancing at the paper. One thing she doesn't understand—it is Monday night and he doesn't have the ball game on!

A week later she consents to go out for dinner. Perhaps he's not all bad after all. He *has* been spending a lot of time with her lately, even without her really encouraging him. The strange thing is that she believes he is actually enjoying the time with her! As they drive home from dinner she keeps thinking how excellent the seafood was—this being the first time he has taken her to a seafood restaurant.

Another month passes and he suggests that he needs to get away for a few days from his work, that he doesn't care where they go—he just wants to get away and do nothing. Does she want to run down to the beach for a long weekend?

*

The Underlying Cancer In A Poor Marriage Is SELF

*

"It's been three months since I told him I didn't love him anymore," she thinks to herself. "It didn't seem to phase him that night I told him. He simply got up and went to work the next morning like nothing had ever happened. He has not mentioned my statements since then. I'm sure he heard what I said, but I'm glad he didn't make me chisel it in stone over and over again. I'm also glad he didn't voice my shortcomings to me. I still feel he is 90 percent to blame, but I'd just as soon not hear his viewpoint on that 10 percent of my faults. It's better now than it was then, but I wonder if it can last. Perhaps there is some hope for my world to not fall apart." She lets her mind wander. Can there really be hope for their marriage? Has her husband really been changing over the past three months? She is not as down all the time as she had been. She is not, however, going to be let down again as she has been before. She is going to keep her guard up. Is he trying to buy her back? Is he sincere or just accommodating? She has been hurt so much in the past.

She declines the weekend at the beach.

The husband is hurt that she doesn't accept, but understands why: He hasn't yet built up her trust in him. Yet he is not going to let that slow him down because he has set his goals and is determined to persist in them until they are reached. It's been three months since his last visit to a full-potential husband, and he's ready for the next lesson. He has not arrived yet, but he's getting closer. He wants that marriage which is planted in his mind more than anything else in the world. He wants to know what the next session will bring.

The wife begins to look at herself differently—not a lot, but just enough to begin asking some questions about her responsibility in marriage. She had confided in some friends who invited her to a Bible study group where the topic was centered around the wife's role in marriage. They discussed such things as "Submission to your husband: Is it possible today?" "Understanding the needs of a husband," "The quarrel, who really wins?" and "Is divorce an option?"

She knows about her responsibility but feels she has taken enough. She has tried to make the marriage work, but he does too many things that hurt. If he would just do his part, she would do her part. She almost wishes she had agreed to go to the beach. She admits she is being selfish but lets her decision stay.

About the most stabilizing process that has taken place that he has started is the time they spend together at night just before going to bed. He has been very adamant about this time together and she is beginning to feel a faint unity developing between her and her husband. The women at the Bible study made a suggestion to her when she told them what a difficult time she has to even consider praying out loud with her husband. They suggested that she and her husband give prayer requests to each other for prayer. This has been a big help. They each are feeling more of a togetherness when they bring up situations with the children, the work, problems that come up every day. She actually looks forward to that time together because she knows he is sincerely interested in her and the family by the way he asks God for direction as a father and as a husband. He tells God some things he is thankful for about her that she didn't even realize he noticed about her.

She believes things are going in the right direction in their marriage but there is so much that has to happen before it is solid.

The Secret of Loving Your Wife

THE man makes a phone call to the next full-potential husband on his list, and in less than a week he is entering the office of the man who will teach him more about what it takes to insure a happy marriage. He finally realizes that the foundation of how he does his job, how he looks at other people, how his entire life will be directed, depends mainly on how he handles his relationship with his wife. If he can't run his marriage successfully, how can he expect to run his business successfully?

As he enters the office, he notices a difference between this full-potential husband and the last one. There is no dark suit, no tie, no starched collar. Instead, Mr. James Clement is rather laid-back, with a new cotton knit shirt, neatly pressed khaki pants, and a nice bronze tan. Mr. Clement is in business for himself, in sales, and his wife works with him.

"So you've met other full-potential husbands already?" He appears so relaxed and in control.

"Yes, I've met Mr. Commitment and Mr. Wife-Time-Investment," laughs the man.

"Then you've learned how to build a good foundation for a tremendous marriage relationship," replies Mr. Clement.

"Just remember," he continues, "if you aren't having a great time with your wife, a large percentage of your life is an unhappy time, which carries over into everything you do."

"That's why I'm here. I want to reach my full potential in life, and I realize I must first reach my full potential as a husband. I want a happy marriage."

The man notices a picture of Mr. Clement's wife on his desk.

"That picture is a reminder to me," the full-potential husband responds. "It reminds me to say 'I love you' every time I look at it.

"You see, I'm supposed to talk to you about loving your wife, but this love begins in your mind long before it ever shows. That picture helps remind me to spend some time thinking how much my wife means to me.

"Do you read the Bible much?"

"Some."

"Proverbs 23:7 says, 'As a man thinks in his heart, so is he.' To me this says that whatever you put into your heart will sooner or later show up as the real you. You become what you think about all day. Your thoughts about your wife will sooner or later become evident to her. The secret to becoming a full-potential husband lies in what you feed your mind about your wife. I think this is the most important fact you have to learn in order to develop your goal about your wife. I stress that last statement: The most important fact you have to learn is to repeatedly feed the right thoughts about your wife into your mind."

"What you're saying," the man responds, "is that you have to *think about* loving your wife before you can really love her."

*

True Love Builds HER Up

*

"Right. First you have to feel it, and then it shows. I want to teach you how to develop love for your wife. Do you remember how often you thought about your wife when you first began to date her?"

"Very, very often," the man responds.

"Did you concentrate on her good points or the bad ones?"

"I don't think she had any bad ones when I first started going with her."

"Sure she did; she had them and you knew it. You just put the bad points in the balance and weighed them against her good points. Your love for her was so pure that it greatly outweighed anything she did which would make you not love her. True?"

"I guess that's a fair statement."

"Did you work hard at getting her to decide that you were the one for her? Did you do things for her and to her that you don't do today?"

"Well, sure. I worked at it a good deal," the man responds. "But it's different now. I don't hold her hand very much or open the door to the car. Things like that. She doesn't expect that now—not after all this time."

"When you showed your deep love for her in so many ways, did she respond?"

*

Working At A
Great Marriage
Is An
Everyday Job

*

"We didn't argue very much. I suppose I accepted a lot of little things she did without getting upset at them. It was good back then."

"Would you like it to be good like that again?"

"Sure, who wouldn't?"

"Then tell me, what could you do, like you did back then, to redevelop that love relationship, to rekindle that exciting glow that once was there?" The full-potential husband lays the question before the man, then looks him in the eye and waits. A long silence prevails, but he just waits.

"I haven't worked at keeping that glow going like I did to get it started. I haven't worked at it, but I would like to. Can you help me?" The man has an eager heart and is ready to take notes on what he should do. "Would you go into more depth on what you were saying about putting thoughts into my mind? You talk like it is almost a secret cornerstone of the foundation."

"Yes, it is almost like a secret—one that most men either don't realize or else overlook. That secret lies in the heart. You must develop a love for her in your heart. Let me read you that verse from Proverbs in the Bible which says, 'As a man thinks, so is he.' Emerson said it differently: 'A man becomes what he thinks about all day.' It's true. If you want to develop a certain relationship with your wife, just begin developing it in your mind right now. Keep it in front of you at all times, think about it all day, and that relationship will slowly begin to develop, until one day it becomes a reality. The secret cornerstone you just asked about is fairly simple: Think about your wife often, praise her in your mind, and keep her quality actions in the forefront of your mind. Soon you'll have a love for her that has been rekindled to the magnitude of those original flames. Remember your love for her back then at the beginning? Well, the great part about becoming a full-potential husband lies in the fact that your relationship not only becomes as great as it was initially, but it continues to increase and multiplies 20 times over."

"That would be great!" the man responds, excited at the possibility of a marriage and a life he had once dreamed of.

"It all begins in your mind. Remind yourself that you love her as often as possible. Put reminders of her where you will see them throughout the day," Mr. Clement comments.

"Like reminding yourself you love her when you look at her picture on the desk?"

"Right. Whenever I think of her during the day, I think of the good qualities, of the positive side of love, of what she does for me and the family."

"That sounds good, but what about the times when she doesn't hold up her end of the love relationship? It's hard for me not to think of those times when I think about my wife. Your wife may be a lot different from mine."

"What exactly do you mean when you say 'not hold up her end of the relationship'?" The full-potential husband begins to set the stage for what he wants to say.

"Well, I think marriage is a 50-50 proposition. I was quarterback in high school and my wife was a cheerleader. I took her back to the football field to propose to her—on the 50-yard line. And I proposed to her on the 50-yard line for a good reason. That may sound a little corny to you, but I did it to emphasize the fact that marriage is a 50-50 proposition. I have half the responsibility, she has half. My question is, How do you love your wife in your heart when she doesn't do her 50 percent of the relationship?"

"That's an interesting concept, that 50/50 deal you speak of. But it has one problem."

"What?"

"It doesn't hold water. Want to know why?"

"Sure."

"Example: My wife and I had an argument." Mr. Clement has been very relaxed until now, but stands and walks directly in front of the man so they are face-to-face. "She was wrong, dead wrong. I was right—no doubt about it. Get the picture?"

"You're sure she was wrong?"

"Positive. Even she admitted it the next day.

"She cried, then went storming out of the room and off to bed."

"I know the scene; you don't have to draw me pictures!"

"For the first time in my life, I asked the Lord how to handle the situation. I had been a Christian but never really thought about getting God in on the everyday scene—like, what would You direct me to do right now in this particular situation?"

"What happened? Did you hear voices?"

"No voices. But He did speak to me through His written Word, the Bible."

"How?" The man is becoming more and more interested as the story unfolds.

"I don't know why, but I turned to Ephesians 5, where Paul was giving instructions on how Christian husbands and wives are to relate to each other."

"What did he say?"

"One thing which said it all to me: 'Husbands, love your wives...' I looked at that verse a long time. I looked at the verses in front of it and the ones behind it. Surely these verses were instructions. As I read, I asked, What if she is definitely wrong and I am definitely right?

'Husbands, love your wives.'

"What if she doesn't uphold her side of the deal?

'Husbands, love your wives.'

"Does that mean I have to love my wife even in a situation where I win and she knows it?

'Husbands, love your wives.' That was the only definite instruction that this passage was telling me: to love my wife, period."

The man interrupts the full-potential husband: "What did you do?"

"I woke her up, read her the passage, and told her I was going to love her from then on because God had directed me to."

"Even if she didn't cover her 50 yards?"

"Even then. You see, it's not a 50-50 proposition. That Scripture was telling me that my responsibility is 100 percent. It's a 100-100 proposition. The 100-percent effort on the husband's part has to be independent of the wife's effort, whether her effort is 100 percent or 10 percent."

"I see. I need to think loving thoughts about my wife whether she is doing her part or not."

"Exactly. God's love is an unconditional love. Man cannot love unconditionally. Sooner or later man's love becomes conditional. The only way this plan works is to let God love your wife through you. This gets self out of the way and allows a love to be expressed that supersedes any love you or I could ever have toward our wives.

"But do you know the most exciting part of this whole session?" Mr. Clement walks back to his chair, leans back, relaxes, and crosses his legs.

"I'm listening—what happens?"

"Pretty soon you have built up your wife in your mind. She again becomes the girl of your dreams. She responds. You begin to build her up, not only in your mind but outwardly. You can't keep what's in your mind a secret. What you feel about her will eventually begin to show."

"Sounds good. I think it could really work."

"It *does* work. I assure you, it really does work."

"Okay. I can handle this idea that I must work on thought time about the positive side of my wife, that in order to love her outwardly I must first love her inwardly. I will commit to that. What next?"

"The next step," begins the full-potential husband, "is a natural next step. If you really love someone deeply, what are you going to do?"

"I'm going to tell her," responds the man.

"Of course you are. Now comes the difference between 'I love you' and 'I REALLY love you.' "

The full-potential husband has it down so pat, so detailed that there is no doubt in the man's mind that he is speaking to an authority.

"And what is the difference between saying 'I love you' and 'I really love you'?" the man wants to know.

"The difference is twofold: detail and repetition."

"What do you mean—detail?"

"Well, you don't just say 'I love you.' You verbalize in detail what it is that you love about her: the way she dresses, the time she spends with the children. You love the fact that she cooks for you; you love her for looking after your clothes. You express in detail those things you have been thinking positively about her all day."

"I understand what you mean about the detail; and the repetition enforces what you think." The man is beginning to understand what the full-potential husband has been trying to get across to him all morning.

"One hundred percent correct. Tell me what importance you think repetition plays in expressing your love to her." Mr. Clement smiles and decides he is being successful in conveying his thoughts.

"If I really praise her and express my love to her repeatedly, it will sooner or later make her realize I really do love her."

"Correct. And what do you think will be the outcome?"

"Once she realizes that my love is 100 percent real, that I have worked it through my heart, that I sincerely love her more than anything or anyone else in the world, then she will respond to that love."

"You're catching on. That's the way it works over the long haul. There will be times when you may think there is no response, but in the end it will be there."

"Why didn't I think of that? Why have I always protected myself against the fear of being hurt by no response? I'm ready to give it my best," responds the man.

"Don't you want to know the final key to the marriage you're looking for?"

"I sure do. What is it? What's the next step in developing this love for my wife?"

"It's one of God's greatest plans—an essential part. Without it a marriage will come apart at the seams."

"What is it?"

"You must physically love her."

"Oh, I know that. That's no exclusive secret you have."

"What I am talking about is something deeper than that. Now, don't get me wrong. The act of sex is important. Very, very important. As a matter of fact, I will make a statement which is at least 99 percent correct: I have never seen an extremely great marriage where the sexual aspect was not extremely great. If your marriage is missing out on that, you have the strongest indicator of big trouble."

"I hate to admit it, but that's where we are right now."

"That makes it a tough problem to solve, but not an insurmountable one. It means you have to be even more committed to becoming a full-potential husband. It will happen in time."

"How can you be so sure?"

"Because that is God's plan for husband and wife and we are basing marriage on God's plan as we find it written in His instruction book: the Bible."

"However, I'm not just talking about going to bed with your wife. God's plan comes naturally when His love is in your heart. His plan includes hugging your wife when you get home. His plan includes letting your children catch you standing at the kitchen sink embracing each other, it includes walking around the yard with your arm around her shoulder or waist, it includes touching and physical contact with each other."

"I don't do much of all that you're mentioning, but I will. I understand what you're saying."

"When you reach this point in loving your wife, the Scripture about love begins to come alive and gives us a pattern to live by."

"What Scripture?"

"First Corinthians 13:4-7. Let me review it for you. It's a difficult dose of medicine that becomes easy to swallow as you work toward becoming a full-potential husband.

"It says that love is patient and love is kind. These two go hand-in-hand. If you are patient with your wife and truly want to love her, you will end up by being kind to her even when she is not doing her 100 percent."

"Makes sense. Go on." The man takes notes.

"It says that love does not boast and is not proud."

"I get it," responds the man. "If I'm boastful and proud to her, I'm trying to build myself up to her rather than trying to build her up."

*

Don't Waste
A Good Tomorrow
On Differences
That Happened
Today

*

"That's right. If you will notice, a strong leader always tries to build up those around him. And they will respond to that leadership.

"Let's go on." Mr. clement reads on out of the Bible. "Love is not rude, is not self-seeking, is not easily provoked, keeps no record of wrongs."

"I usually keep records of my wife's wrongs for about four days," laughs the man, "and those are a terrible four days. You're right. It's during those days that I'm rude to her; I get angered by her very easily. She can't do anything right during those times. I'm keeping record of what she has done to make me angry. This love you're talking about doesn't keep any record of her wrongs? It can't even carry over till the next day?"

"Not even until that night. Settle differences before you go to sleep. It's no use wasting a good tomorrow on differences that happened today." Mr. Clement looks at the man and pauses a moment to let it all sink in.

He reads on: "Love always protects, always believes, always hopes, always perseveres."

"Does that mean," the man interrupts, "that part about 'always protects,' that I can never talk my wife down to my friends when she really does something that upsets me?"

"If you talk disrespectfully and distrustingly about your wife to one person, do you honestly believe that it will stop there?" Mr. Clement leans forward and waits for a reply.

"No. He'll tell his wife and she'll tell her best friend, who will discuss it at the next bridge club. You're right—that's not a very protective husband. It's like I've given up hope in her for the time being. That Scripture you just read says 'Love always hopes.' " The man nods in knowing agreement.

"Good insight," Mr. Clement says. "But remember, it's not *I* saying it, it's the Scripture saying it.

"Look at the next statement. It's a profound statement that is both factual and a promise: 'Love never fails.' The kind of love we must develop as husbands is one that never fails our wives. It is one they can depend on, no matter what."

"And a promise?" The man asks for further explanation.

"And a promise: If you offer that type of love to your wife, it will never let you down, it will not fail you."

"Great!" responds the man. "I never looked at being able to love my wife quite that way."

Mr. Clement feels satisfied with the session. He knows that the man now understands how to develop a true love for his wife. He is ready to let him go on to the next step in becoming a full-potential husband.

"Do you mind if I review my notes with you?" the man asks.

"Not at all," the full-potential husband says. "It will let me see how good a teacher I've been."

"To become a full-potential husband, one must consciously work at loving his wife.

"Love begins in the heart before it is ever shown in action. Think only on the positives of your wife.

"Marriage is a 100-100 proposition. The husband is responsible for his 100 percent. Husbands, love your wives—period.

"Verbalize your love to your wife. Tell her that you REALLY love her. Tell her in detail and with repetition.

"Physically love your wife."

The review completed, the man thanks the full-potential husband and heads home to continue in his growth toward his new goal of loving his wife.

He stops by his office on the way home from the airport. The first thing he does is write Ephesians 5:25 on a small piece of paper: "Husbands, love your wives." He slides the paper under the glass on his desktop so it will be in full view to him as he sits at his desk. He also picks a spot on the wall where he will hang a picture of his wife—another reminder to think of her throughout the day. He next opens a drawer and shuffles through some papers, looking for something his wife gave him when they were married. It is a pen desk set with their anniversary date on it. He didn't like the pen in the set so he had simply stuck the set in the drawer, still in the original box. She has never mentioned a word to him about it not being on his desk. He removes the white tissue paper from around the walnut base and places it squarely in front of him, easy to reach from his chair. It will now remind him of days gone by, when he looked at her and loved her differently from the way he does now. He will begin to use the pen just to remind himself of his wife.

The husband leaves his office and starts home. He has some time to think about how much he really does love his wife. He sees how important she is to him in his daily life: her smiles; running the family; taking care of his clothes; accepting his shortcomings. He really does love her—there is no doubt in his mind. He has mishandled the marriage so many times, but now he is going to approach it differently—scripturally.

For the first time in her marriage, the wife notices something very different. She begins to see a humbler man as her husband. It is not so important for him to make sure she knows when he is right and she is wrong. She notices that he hangs around the kitchen when he comes home. He is actually not much help, but it is nice for him to be around. He drops a lot of ice cubes on the floor, breaks one of the good glasses, and somehow snaps the handle off a coffee cup, but she likes his presence anyway because it has eased the tension more than anything they have done together for years.

He rarely watches television anymore. He took her to get ice cream last night. The hatred she once had in her heart is melting. . .slowly. His touching her does not seem nearly so awkward as it did just a few months back. It's strange to say, but he is acting almost like he did when they first started dating. The wife believes he is actually trying to be a husband. She is beginning to look forward to his getting home in the evening.

She doesn't know what has happened, but whatever it is, it's real. There is a certain consistency to her husband that is great. She actually believes he may love her after all. She just hopes that their relationship will continue to improve. She has been praying for him, and she has been praying for the Lord to give her guidance also. She never wanted a divorce anyway; she always wanted a happy marriage, and now she thinks there may be a glimmer of hope. She will wait and continue to pray.

Their sexual relationship in the marriage is changing. She is actually beginning to feel some sort of desire for him again. Not much of a change but a far cry from several months ago when she couldn't stand for him to touch her.

She has continued her Bible study class with her friends, studying the role she is supposed to play as a Christian wife. She had difficulty when they studied a passage in the Bible she had never paid much attention to before. It is found in Ephesians 5 and deals with what a wife is to do. At first she had trouble with the part about "submitting to your husband," but she is beginning to realize that it doesn't mean to submit like a slave to a master but like a musician to a conductor. She studies what submission under God's direction can do to influence a husband.

The other women in the class pray for her. For the first time since she told him she was through, her heart begins to change. One part of her tells herself that she hates him but now another part is telling her that God has a plan for marriage, a plan she has not followed before. "Self" tells her it can never work: He is too rotten and self-centered, and it's not worth trying anymore. She is confused. She wants to believe what the Bible instructs her to do, but deep down inside she doesn't really believe that her particular situation is amenable to what Scripture is saying. She promises God only one thing: She will pray about what responsibility she has in the marriage.

Essentials of Leadership

THE man's next flight takes him all the way across the U.S. to the office of the chief executive officer of a large corporation. He wonders how this full-potential husband relates his goals of work to those of becoming a full-potential husband. He is soon to find out as he waits in the plush waiting area outside the office.

At exactly ten o'clock the full-potential husband, a Mr. Jake Shaw, walks out to greet him. "It's good to actually meet you," Mr. Shaw says as he ushers him into his office. "I've heard how receptive you've been to the others in the club." He offers the man a seat.

The man glances around the office. He notices a small plaque on the desk that reads "LEAD, FOLLOW, OR GET OUT OF THE WAY." That should give him a clue to what they'll discuss, but he's not sure what it has to do with being a full-potential husband.

"If you don't mind, let's get straight to the point of the matter. I've been placed at the head of this organization," Mr. Shaw begins. "In any organization, great or small, there has to be a leader. If I'm not a full-potential leader, the whole structure of the organization hurts. I have to make decisions every day, and the success of this company depends on whether these decisions are correct.

"If a decision concerning the company is one I dislike personally but is good for the company, I must forgo myself and decide in favor of the company. Likewise, if some of the officers in the company favor something I know will not be in the best interest of the company, I have to take a stand against them and express my authority as chief executive officer."

"But don't you make the wrong choice sometimes?"

"Occasionally, yes. That's why I have to give such decisions so much consideration. That's the responsibility assigned to me by the board, and I must accept that responsibility.

"That's what I want to talk to you about—being the head of your family. You have the God-given authority to direct your family toward becoming a godly family."

"What do you mean, God-given authority?" the man asks.

"I happen to believe in the Bible. I believe that it is as directive today as it was when it was written. I believe its instructions about husband and wife in Ephesians 5. You can read that the husband is the head of the wife just as Christ is the head of the church."

"That sounds mighty authoritative to me, and I know it would sound authoritative to my wife as well." The man doesn't buy the idea at its face value; he knows his wife won't see it that way either.

"I knew you would react to the idea just the way you're doing. You don't understand it now any more than I did when I first read it," Mr. Shaw explains.

"You're right; if that's a workable situation in today's world, there must be something I don't understand." The man stands firm in his belief that if he stresses the fact that he is the head of his wife, it is not going to help at all. In fact, it will probably make matters worse.

"The passage says that the husband is the head of the wife just as Christ is head of the church. Do you know how much Christ wants the church to succeed today? Do you know how much Christ loves the church today?"

"Go ahead. What are you getting at?"

"Christ loves the church enough that He was willing to die for it. Can you understand that? Do you think for a minute He would do anything to hurt that body called the church, that same church He went through scourging for, where he was flogged with a whip having lead balls and sheep bones tied into leather thongs until he was just short of collapse or death? The whipping was so severe that He couldn't even carry His cross. Do you understand what kind of love that was and is?"

"I never thought of Jesus' love in that light before," responds the man.

"Well, just think about how much a husband is supposed to love his wife: *as Christ loved the church*. Just think about the suffering He went through to prove that love. Just concentrate on the flogging for a minute, and forget the actual crucifixion. Can you imagine Him with those bleeding wounds in His back muscles and insects lighting on the open wounds? Concentrate on the intense pain and suffering He was going through. Then He gets to the site of the crucifixion and they threw Him to the ground on His back to nail His hands to the crossbar. Can you imagine what He was going through just because He loved the people who were to follow Him in days and years to come?"

"Never thought of it in that detail before," answers the man.

"You have to think about it to understand this passage. You are to love your wife in the same way. Jesus Christ did all this for you and me even though He knew how we were going to treat Him. And you think you are some kind of hero if you forgive your wife for some little incident. We as husbands are to love our wives in the same way Christ loved the church, and that church is every believer who ever lived." Mr. Shaw has made a strong point.

He continues. "Do you think anyone who loves something that much would ever treat it in any way other than to build it up, make it stronger, make it better in every way possible? When you believe in something enough to die for it, you are taking on a lot of responsibility for it. Agreed?"

"I agree with what you're saying so far."

"If you loved something that strongly, you would love it, care for it, and nourish it. If you saw anything that would remotely threaten to destroy it, you would authoritatively step in and do whatever you could to save it."

"But what does that have to do with my telling my wife she has to obey my wishes and whims, that I am head over her?" The man keeps bringing the matter back to the reality of his home.

"That's the secret. What you just said is the secret to understanding this passage of Scripture." Mr. Shaw acts almost gleeful that the man has made that point about every wish and whim. "If you love your wife like Christ loved the church, your wishes and whims will not be a factor in your heading the marriage the way it should go any more than I let my wishes and whims influence me in deciding how I run this company. This Scripture puts so much responsibility on you as a husband that you have to be on constant guard to remain responsible to God for the authority He has given you."

"That's awesome." The man begins to realize what his responsibility as a husband is all about. He begins to see that his role as a husband is not an authoritative *freedom* but an authoritative *responsibility*. "I think I understand a little bit. For the first time in my life, I think I see a little glimmer of what you mean when you say that the husband is the head of the wife. I foresee a lot of changes I'm going to have to make about the way I look at myself in charge before it will ever work. I accept your foundation on the authority of a husband. Now tell me how to use that authority of a husband." The man becomes excited with this newfound outlook on his role as husband.

"You see the need for there to be someone at the helm? You realize that even a committee has a single individual as chairman in charge? You understand the necessity of a head of a home, someone to have the final decision, which is not self-seeking but has the unity of the home in mind?" The full-potential husband lays it on the line for the man to see.

"Yes, I realize the need."

"Do you realize the responsibility?"

"Yes, I do."

"Okay, then we can proceed with the types of leadership you can have. There are different types of leadership that a husband has to recognize in order to be an effective leader." The full-potential husband begins his outline.

"*Consideration* leadership is the most impressive type of leadership that any husband can exhibit."

"Consideration leadership. You're going to tell me to be considerate of my wife? I see myself as a fairly considerate husband. I opened the car door for her the other evening."

"Did anyone see you do it?"

"Sure, we were leaving a party and lots of people were around to see me."

"Did you open the door for her when you left home going to the party?"

"No, I don't believe I did."

"No one was around to see you do it, was there?"

"Okay. You made your point. I didn't consider opening her door out of my love for her as much as I did out of show. You have my attention. Go ahead and tell me about the considerate leader." The man readies himself to take notes.

"The considerate-leader husband stays tuned into his wife. He listens to her—not just what she says verbally but also what she says with her actions and her attitudes about certain events and ideas. He listens in love."

"Listen in love—what do you mean by that?" the man interrupts.

"You listen to know where she wants to eat, you listen to what type of movie she likes to see, you listen to where she would like to go on vacation. All wives tell husbands these things, but unless we listen to them with love in our hearts, we tend not to hear what they're saying."

"I see," the man nods. "I begin to acknowledge in my mind what she wants out of life and take that into consideration when we go to a movie, eat out, or go on vacation."

*

The Key
To Successful
Communication
With Your Wife

LISTEN

*

"Exactly."

"So I consider what movies she wants to see. Do I always have to see a love story every time we go to the movies? I hate love stories. I really don't like them."

"You have wife on one side of the balance and self on the other. Do you love her enough to give up self in that situation?" Mr. Shaw makes it difficult for the man.

"Okay. I love her enough to give up self. Does that mean I never get to see a good Western movie?" The man hates to give in on this completely.

"Oh, I'm sure she will respond sooner or later if you ask her if she will go to a Western movie with you. But that's not important. The important point is the fact that you are building your role as a leader. Consideration leadership is a very important part of the overall picture of the husband as an effective leader."

"No big deal," responds the man. "I can accept that role. What's the next step?"

"*Involvement* leadership." The full-potential husband wastes no time; he simply starts the next point.

"In consideration leadership you know without a doubt what you want and what your wife wants, so you consider her side and act accordingly. However, in involvement leadership there is more involved than a movie or where to eat. The stakes are greater and the outcome is more important. Perhaps you don't know her side of the decision. Perhaps you don't even realize she has an opinion about a situation."

"So what do you do?" the man asks.

"You get her involved in the decision-making process. In involvement leadership you ask your wife's opinion. You discuss it, then you make your decision as a leader.

"That's what I do as chief executive officer of this corporation. With some decisions I involve many people, but I'm still the leader. I still have the responsibility even though I involve others in the decision-making on some matters."

"What's an example of involvement leadership with your wife?" The man asks for specifics.

"I looked for some land to build a house on. I was set on a piece of land with a beautiful view. But that's all it had—a view. It had very little usable land. My wife wanted a piece of land that had usability to it—a yard, a drive, some space around the house. Before I decided on a building site, I conferred with her. She pointed out specifics about why the farm with usable land would be better than the farm with the dramatic view, when considering our particular situation. After talking with her, I realized that I had tunnel vision and had not properly appraised all the negatives in my choice. It was not that she was all right and I was all wrong. She had no concept of the prices of the land or the value of the land. It was still my responsibility to decide which was the best value. After taking everything into consideration, after getting to the bottom line, we bought the farm with the usable land."

"Wasn't it hard to admit you were wrong?" the man asks.

"It wasn't a matter of right or wrong. It was a matter of which was best for our family and our lifestyle, and then doing it. Remember, a successful leader has to take self out of the picture in order to be objective in his decisions." Mr. Shaw makes his point. He has shown the man that a leader of an organization has to rise above his own personal wants and wishes whether that organization is a corporation or a marriage.

"I'm beginning to get the real feel of what a full-potential husband is all about." The man begins to see the magnitude of working toward his goal as a husband. "You're asking a lot when you start talking about removing self from the picture and looking at a situation objectively. That sounds almost impossible."

"That's the key." Mr. Shaw stands and walks over to the window for emphasis. "It really is impossible if you try to do it yourself. Man is made that way. We are basically selfish creatures."

"Then how can it work?"

"It can only work by asking God to rule over our hearts rather than self. We must stop and ask Him for direction in times of decision-making. That way we will be yielding to His decision in our marriage rather than self's direction. After all, He ordained marriage and He's going to instruct you in the proper direction to make it work."

The man says nothing. He only stares at his notes. He thinks back over the numerous decisions he has made about his marriage. "I can't ever remember praying about a specific decision I've ever made concerning day-to-day matters of my marriage. Did you really pray about which land to buy?"

"Sure. Just the act of turning something over to the Lord helps get self out of the picture. Once that happens, you look at every situation differently."

"That's overwhelming to me," the man says. "How many times do you pray about everyday events in your life?"

"If you include my marriage, my business, and my personal life, probably 20 times before noon." Mr. Shaw sits back down. He has made a profound impact on the man, a soul-searching impact which is going to affect him for the rest of his life. He has always considered himself a Christian but has never thought about how it could affect his everyday life. . .20 times before noon!

"What you're saying, is that to have the marriage I want, I have to change myself rather than trying so hard to change my wife."

"Exactly. You can't change your wife, but God can change her through you if you will continually ask His direction."

"Okay, if you say it works," the man responds, but with a hint of skepticism. "But I see one big flaw in all of this."

"And what's that?"

"What if the decision I reach goes against my wife's wishes. Then what?"

"Good insight. That brings us to the last type of leadership: *full-responsibility leadership.*" The full-potential husband hits the man's concern head-on.

*

The Best Way
To Have A
Great Marriage
Is To
Become A
Great Husband

*

"The full-responsibility leadership husband is one who has gained the respect of his wife through his leadership as we have already discussed. The first two steps, consideration leadership and involvement leadership, act as foundation stones for responsibility leadership. If the husband has been sincerely building the marriage relationship based on Scripture, then, when the time comes that he must make a decision on which he has to take a stand alone, he will take full responsibility for that decision. There are some situations you must control, some decisions on which you will have to take a firm stand as head of the family. That stand may bring a boiling-point situation in your marriage, but it must be taken." Mr. Shaw stands firm in his instructions.

"But what if she doesn't accept your authority?" The man is uncertain that he has that type of leadership.

"First of all, if full-responsibility leadership doesn't work, it's because you haven't developed through the first two leadership phrases. You can't just *demand* respect; you have to develop it through love for your wife. It has to show through each and every decision you make as a full-potential husband. When the time comes for you to take the full responsibility for a decision, the responsibility-leadership rule will automatically come into play. She will honor your decision."

"What if your decision is wrong after you take this strong stand?" the man persists in his questioning.

"You have to be very careful when using full-responsibility leadership because of the question you're asking. If it is ever used wrongly, it destroys much of the effectiveness of the leader. I use this as little as I can, and I always turn to prayer and the Scriptures for this type of leadership response. But there are times when you will have to use it. There will be direct decisions which you must control for the unity of the marriage. Do you think you can handle it?"

"I'll have to work at it," the man responds, as if accepting an insurmountable task assigned to him. "As I see it, your most significant point was when you stated that there had to be a building-up process in developing a leadership role. I have a long way to go, but I feel sure I can do it."

"Just remember these qualities of good leadership and you'll develop into a good leader. Recognition is important," Mr. Shaw continues. "Recognize what your wife does for you and the family. Recognize her importance. Usually when a wife does something good, the husband doesn't say anything. When she makes a blunder, he tells her about it right now. It doesn't take many times like that and soon her commitment to marriage becomes less and less, until one day she lets the thought of divorce creep into her mind. Soon she has a 'don't-care' attitude and loses more interest in the marriage."

"I know that feeling," responds the man in agreement.

"Recognize her for her worth to you. Recognition of the wife is a key to leadership as a husband."

"I've got that imbedded in my mind now. Is there a counterpart to recognition?" The man is excited now. He realizes he is almost there. He knows what he must do to become a full-potential husband. He has made a significant change in his outlook about being a husband. He knows that *changed attitudes* are the secret to *changing actions*. He realizes that a person cannot permanently hide feelings—that sooner or later how you feel toward someone becomes manifested by your actions. His present attitude toward his wife is great. He wants to know how to make it even greater.

"The counterpart to recognition is encouragement. A full-potential husband as a leader is his wife's greatest source of encouragement. How could you encourage your wife in a practical day-to-day way?" Mr. Shaw asks the man to begin putting this into practice in his mind, because he knows that if he practices it in his mind, he will eventually practice it in reality.

"Probably the best encouragement is in the form of compliments," the man responds.

"Excellent. You're starting to think like a full-potential husband. How will you go about doing that?"

"As a full-potential husband, I will start thinking about my wife as I return home on the plane today. I will think of all the things she does that I can compliment her on. She's a wonderful cook, has a nice smile, dresses smartly, even picks up my dirty socks when I forget to pick them up. The list could go on and on."

"First in your mind, then action. Right?"

"Right. The other part will be easy for me. The difficult part is thinking it all through in my mind." The man begins to understand how he usually approaches situations.

*

It's Smart To Love Your Wife

*

"You have finally understood the key to the full-potential husband. You first need to desire to become the best possible husband you can. Set that as a major goal. Then you have to begin working at it consistently in your mind. Finally, action takes place. You can't hide it, you can't stop it—it will happen," Mr. Shaw concludes.

"Before I leave, let me quickly review with you the highlights of what you've taught me."

"Sure, go ahead and I'll check you."

"There has to be one person to act as a responsible leader of a marriage."

"Correct. Otherwise you become a committee with no one leading it."

"The husband is the head of the wife as Christ is the head of the church. He loved the church enough to die for it."

"Go ahead. That puts it in proper perspective."

"There are three types of leadership I must develop if I'm going to accept that responsibility."

"What are they?"

"Consideration leadership. I consider what I already know about my wife's desires."

"Go on."

"Involvement leadership. I involve her directly with the matter to be decided. I place the institution of marriage above self, above what I want personally."

"That's good insight to good leadership. What's the third step?" Mr. Shaw feels that the man has the desire to make it happen.

"The last part is the tough one, the one where I have to weigh all the factors and come up with an answer and then stand firm, taking full responsibility for the outcome. Full-responsibility leadership is what you called it."

"You're right. It's tough, but essential." The full-potential husband puts his arm around the man's shoulder as they walk toward the office door. He feels a certain excitement of satisfaction as the man leaves. He has accomplished something which goes beyond being chief executive officer.

"I thank you so much for your time and instructions. I'll forever be indebted to you. As a matter of fact, so will my wife."

The man heads home and contemplates his newly realized rules of leadership. He begins practicing in his mind even before he reaches home. He thinks back to a previous session. He recalls that his mind will produce whatever is planted in it. He recalls that whatever his thoughts are today will sometime in the future manifest themselves into actual occurrences—that his thoughts about his wife will not remain a secret. They will sooner or later show in his actions and attitudes, whether the thoughts are positive or negative. He now realizes that if he is ever going to be able to show true intimacy to his wife, he must first develop it in his own mind; and if he persists at it long enough, it can't help but result in actuality.

He also thinks back to the goal-setting session. He now realizes that if these thoughts about his wife are not tied directly to a purpose, a goal, nothing will be accomplished. He needs to set goals directed toward being a better husband or else his thoughts will drift around aimlessly, with no order or direction to them.

He commits to his goal of becoming a full-potential husband under God's direction. He will keep that goal directly in front of himself mentally, looking neither to the right nor the left, until he is face-to-face with the successful accomplishment of that goal. He will sacrifice whatever it takes. He is determined to not let the advice of these full-potential husbands be in vain.

The wife has changed over the past year also. She attributes much of her change to the group of women in the Bible study she joined. She has had to learn what it means to submit to her husband. She is reminded of the statement that this type of submission is not like a slave to his master but like a musician to his conductor.

Probably the one change in their life which has pulled them back together has been the time they spend together in reading the Bible and praying. Life has become so much more real when they pray about the children, or his job, or some concern they share. Always before it has been so awkward for her to pray in front of her husband, but now it is so natural.

Neither the husband nor the wife knows who was primarily responsible for pulling the marriage back together, but one thing is certain in her mind: Though she used to hate him before, now she loves him. She gets excited now when she sees him drive into the driveway after work. She thinks back to the days when she was ready for it to be all over. It scares her to even imagine how different her life would be right now if she had gone through with a divorce.

She thanks God for giving her a foundation to rebuild on. She thanks Him for a love that only could come from His instructions found in the Bible.

"Do you know that knoll on top of Rich Mountain where you can see for miles in every direction?" The man catches his wife off-guard this afternoon. The children are all at friends' houses, and she knows that she and her husband were going out to eat tonight, but she didn't expect him home so early in the afternoon.

"Sure, I know exactly where you're talking about. We had a picnic up there years ago. Beautiful place. What about it?"

"Let's go up there for a little while before we go out to eat."

"I'm really not dressed to go out. I need to finish cleaning up the kitchen before I go anywhere. Let's just plan to go another time."

"I have a surprise for you. Come on now—just get your sweater and lock the house." The husband persuades his wife to go now. He thinks he has developed into a pretty good leader at times like this. For whatever reason, his wife consents to go.

It is only a ten-mile drive from their home. They park near the top of the knoll, and as they get out of the car and walk to the hilltop together he pulls some papers out of his jacket pocket.

"The time has come for us to begin a great experience." The husband has thought this event out in his mind and has prayed about it in his heart a thousand times during the past few months. "I am a different husband today than I was a little over a year ago, when you told me you were through with me. You don't know what I've done to improve our relationship—the help I've received, the goals I've set, the time I've spent praying about our situation. Whether you know it or not, you are married to a different husband now than on our wedding day. I wish I had known then what being a husband was all about. I didn't appreciate what I had in you as a wife, and I didn't acknowledge the times when I thought you were a great wife." The husband talks to his wife but can't quite seem to look her in the eye; he keeps gazing at the beautiful scenery around him.

She doesn't care whether he looks at her or not. This is the best conversation she has ever listened to. All she can do is stare at his lips and watch the most sincere expression she has ever seen on her husband's face.

He continues. "What I'm trying to say is that I have always loved you but I took marriage for granted. I thought that after we were married I would work and you would support me in all I did. I never gave a second thought about how important you were to me. I'm not saying that you've always been the perfect wife, but I am saying that I did very little to enhance you to become a good wife.

"I've spent the past year trying to improve the part of our marriage that I can be directly responsible for. I may have left you out in the cold by not telling you all I was doing, but I knew I had to get my life straight before I could ever ask you here today."

She is finding it difficult to continue looking at him so closely—not because she doesn't want to, but because it is becoming difficult to see through the tears that for some reason are beginning to well up in her eyes.

"The biggest change I had to make was not one of trying to be a better husband, although that was definitely a needed change. The greatest change was in my spiritual life. I had self running my life and our marriage. The most difficult thing for me to do was to get self out of the picture and then study the Bible and see how God wanted me to play the game of marriage. There were things that I, meaning self, couldn't forgive you for. Then I discovered about a love that could forgive you for all the little and big things I couldn't let go of. That's God's love. His type of love is unconditional. That's how I come to you today. I want God's type of love in our marriage. I have come to realize that the best possible marriage that can ever be achieved is one that is centered around God's teachings."

He finally turns to her. There is a silence which he breaks with a soft smile on his lips. The tense part is over. He has said what he had come to say. Not only is the ice broken, it is melted. Things are natural again. He holds her gently and starts rubbing her shoulder as he remembered doing many times years ago when they would talk about serious matters.

"There is a verse I want to tell you about. It's one I think is very fitting for our situation. Your old husband needs to be forgotten, and we need to look toward a goal which I think God has called us to in our marriage. It's found in Philippians 3 and goes like this: "One thing I do: Forgetting what is behind and straining toward what is ahead, I press on toward the goal, the prize for which God has called me."

"Paul was talking about God calling him to press on in teaching Jesus Christ. I'm talking about pressing on to have the type of marriage which is ordained by God.

"If you're willing to forget whatever may be behind us, and strain toward a great marriage which is before us, I would like to renew our vows right now. I got a copy of our vows from the pastor the other day and have them right here." He holds out a piece of paper, neatly typed and refolded a hundred times where he has reread what he and his wife repeated so many years before. He waits for her response.

"Before I say anything about our vows, let me first say something about your wife you married on our wedding day. She didn't know quite what to expect. No one had told her exactly what role she should play. There were so many things she didn't understand about men in general and in her husband specifically. She too was selfish in many of her ideas and actions. She used the standards of the world, of her friends as a gauge to go by. She didn't know the meaning of a Christ-centered marriage. So your wife of your first marriage needs some forgiving and forgetting also.

"I would like to speak for her and accept that verse you just quoted about forgetting what is past and striving for what lies ahead. Yes, I would like very much to become a new wife today. I would very much like to become your new wife right now."

The man steps back, taking the paper in his left hand and her hand in his right. "Do you realize that there is a different group of people here to witness this occasion?"

"What do you mean?"

"Last time the church was packed with people and I was aware of each one there. This time we aren't even in a church and there certainly aren't any people for miles around. But do you realize that there are thousands of angels looking down on us right now? Do you realize that God Himself is watching? Today we make these vows, and promise before all these witnesses that we will keep them."

"It's awesome when you put it that way, but I'm ready."

The husband begins to read: "Marriage is an institution of God, ordained by God and supported by God. We gather on top of this knoll in the sight of God and in the presence of His angels to join together as man and wife.

"It is my duty to provide for your support, to shelter you from danger, and to cherish for you a manly affection, as commanded in God's Word, that husbands love their wives, even as Christ loved the church and gave His own life for her. With God helping me, I pledge to love, cherish, honor, and protect you, cleaving only and ever to you until God by death shall separate us. I promise this to you."

He kisses her lightly on the cheek and gives some time for thought. Then he looks at the paper once again and reads: "It is the duty of the wife to reverence and obey me, as His Word states that wives be subject unto their own husbands, even as the church is subject to Christ. With God's help, will you love, cherish, honor, and obey me, cleaving only and ever to me until God by death shall separate us?" The husband drops the paper and takes both hands of his wife.

"What do you think?"

"I think," she smiles, "I am the luckiest woman in the world. I will be your wife. I will love you and honor you—even obey you like you once said. Remember? Obey like a musician."

"Yes, I know. Like a musician obeys the conductor who is orchestrating the most beautiful music ever played...our marriage.

"Do you know, I love you?" the husband asks.

"Do you know, I REALLY love you?" she responds.